Klara Marie Ditschkowski

Passing on cultural heritage

Klara Marie Ditschkowski

Passing on cultural heritage

Approaches to teaching Turkish cultural heritage using the example of Kottbusser Tor

ScienciaScripts

Imprint
Any brand names and product names mentioned in this book are subject to trademark, brand or patent protection and are trademarks or registered trademarks of their respective holders. The use of brand names, product names, common names, trade names, product descriptions etc. even without a particular marking in this work is in no way to be construed to mean that such names may be regarded as unrestricted in respect of trademark and brand protection legislation and could thus be used by anyone.

Cover image: www.ingimage.com

This book is a translation from the original published under ISBN 978-620-2-20552-8.

Publisher:
Sciencia Scripts
is a trademark of
Dodo Books Indian Ocean Ltd. and OmniScriptum S.R.L publishing group

120 High Road, East Finchley, London, N2 9ED, United Kingdom
Str. Armeneasca 28/1, office 1, Chisinau MD-2012, Republic of Moldova, Europe
Managing Directors: Ieva Konstantinova, Victoria Ursu
info@omniscriptum.com

Printed at: see last page
ISBN: 978-620-8-98873-9

Table of contents

CHAPTER 1

Introduction

The following chapter provides an overview of the aims of the work. The rough structure and the justification for the choice of space should make it easier to enter the subject area. It also explains which methods of empirical social research were used.

1.1 Aim of the work

This bachelor's thesis deals with Turkish cultural heritage and its communication to other residents in the Kottbusser Tor redevelopment area. The Turkish and German population use the area side by side instead of learning and benefiting from the diversity of the cultures of the other ethnic group. By communicating the rather unknown cultural heritage of the Turks, closer coexistence can be promoted and spatial and social conflict minimised. By showing examples of Turkish cultural heritage in the area, the importance of mediating between two cultures is to be illustrated and the resulting potentials and conflicts explained. Both architectural and spatial cultural heritage will be focused on and comparative analyses of the structures of the oriental city with the structure in the planning area will be presented. Conceptual solutions are developed by combining and further developing existing planning approaches.

The findings will be used to formulate and summarise development strategies to compensate for deficits in the coexistence of two cultures at Kottbusser Tor. The aim of this cultural heritage mediation initiative is not only to promote harmonious coexistence, but also to encourage a targeted exchange of culture and interests.

1.2 Justification of the room selection

Today, the Kottbusser Tor redevelopment area is a vivid illustration of the multiple paradigm shifts in urban planning over the past decades. A radical change of course at the beginning of the 1980s from clear-cut redevelopment to "cautious urban renewal"[1] represents an extraordinary caesura in urban development.[2] The problematic existence of the neighbourhood is particularly evident in the social areas. The area around the Kottbusser Tor has become a centre for foreign workers and their families. As our society often only allowed them to function as labourers in the past, the Turkish residents were hardly able to develop their own identity in connection with their new place of residence and unintentionally contributed to their alienation from the local German population.[3]

Due to the high proportion of immigrant women in the Kottbusser Tor redevelopment area, it is important to include migrant women in the questions and solution scenarios of urban development policy. The aim is to promote coexistence, particularly between the Turkish and German population groups, by communicating the Turkish cultural heritage and thus strengthening identification with the planning area.[4]

1.3 Structure of the work

This section provides an overview of the individual chapters and their

[1] Senate Department for Building and Housing (ed.): Der Block 103 in Berlin-Kreuzberg, Sanierungsgebiet Kottbusser Tor, Berlin 1994, p. 10.

[2] See Senate Department for Urban Development and the Environment (ed.): 23rd Report on Urban Renewal, 01.01.2000 - 31.12.2001, http://www.stadtentwicklung.berlin.de/staedtebau/foerderprogramme/stadterneuer ung/de/download/23StB-00.pdf, p. 53, last accessed 23 August 2014.

[3] See Senator for Building and Housing (ed.): Strategies for Kreuzberg. Project description, information on the project area, objectives. Berlin 1977, p. 1.

[4] See Dame, Thorsten / Herold, Stephanie / Salgo, Andreas (eds.): Re-Vision-IBA '87. 25 Jahre Internationale Bauausstellung Berlin 1987, p. 27.

interrelationships. The work is divided into the following five sections:

- Introduction		(Chapter 1)
- Part 1	Theory	(Chapter 2)
- Part 2	Empiricism	(Chapters 3, 4, 5)
- Part 3	Concept	(Chapters 6, 7, 8)
- Appendix		(chapters 9, 10)

The introduction serves to categorise the subject of the study and to narrow down the topic, as well as to highlight the research interest of the work. Finally, the choice of the research area is explained.

Part 1 presents the results of the basic research. The immigration of the Turkish population to the Federal Republic of Germany is briefly explained. In addition, the structures of the oriental city are presented as an excursus in order to be able to compare them with the urban structures at Kottbusser Tor. Finally, a definition of cultural heritage that is valid for this work is worked out and justified. The context and significance of cultural heritage for urban planning and the resulting potential for conflict are explained.

In Part 2, the presentation of the study area establishes a spatial reference to the neighbourhood . Firstly, the area is presented in its most important development phases, as well as the resulting changes to the building, spatial and social structures. Then, relevant topics from the previous planning approaches of the "Socially Integrative City" and the Integrated Neighbourhood Development Concept (INSEK) are explained and the objectives and measures relevant to this work are presented. Finally, an example of the Turkish cultural heritage in the research area is presented. The focus here is on the architectural and spatial cultural heritage. The possibility of transferring the structure of the oriental city to the urban structures in the planning area is also examined.

In Part 3, an approach to the mediation of Turkish cultural heritage is developed on the basis of the first sections of the thesis. After a description of the necessity, individual key objectives are formulated, which are based on important previous planning approaches and deal with participation processes in the mediation of cultural heritage. After analysing the resulting mediation problems, recommendations for action are formulated and finally a prognosis is drawn up.

The appendix contains the references and the list of illustrations.

1.4 Methodology

Methodologically, the work is based on a sub-spectrum of empirical social research. This means that the complex area can be approached using different methods and the findings are summarised in a coherent overall picture.

In addition to the literature and internet research as well as expert discussions, room inspections and observations complete the knowledge gained. An interdisciplinary literature and internet search was carried out to develop the content. With the help of monographs, bibliographies and online publications, the current state of theoretical research and practical discussion on the principles described above was analysed.

The knowledge gained from expert interviews and site inspections and observations was incorporated into both the analysis of the current situation and the final assessments.

CHAPTER 2

Theory

In the first part of the thesis, the theoretical foundations that are important for the context of the content are developed.

2.1 Turkish migrant women in Germany

At the end of the 1950s, there was a labour shortage in some Western European countries, particularly in the Federal Republic of Germany. This led Turkey, among others, which was suffering from a surplus of labour, to consider sending its surplus workers to Germany. The decisive factor for the pronounced

Fig. 1: Turkish guest workers arriving in Germany in 1960

The reasons for the underemployment in Turkey were the strong population growth, the mechanisation of agriculture and the accelerated industrialisation in the 1950s. Turkey felt connected to the Federal Republic in several respects. On the one hand, West Germany was one of its most important trading and alliance partners. Secondly, the "German-Turkish friendship", which from a Turkish perspective was based primarily on the 'brotherhood in arms[1]during the First World War, played a major role in the consciousness of Turkish women. The German

Foreign Ministry also described the German-Turkish relationship as 'bound by tradition[1].[5]

The recruitment agreement with Turkey was signed on 31 October 1961. Unskilled labourers were mainly recruited to work in Germany for a few years and then return to their home country. Very few of them brought their families with them.[6]

After the economic crisis in Germany in 1973, it was decided to stop recruiting. At that time, there were just over 900,000 Turkish women living in Germany. This would have been a good time to make clear decisions, for example on whether those who had already immigrated should leave the country when their employment contracts expired or whether they should be given the opportunity to stay. However, no clear decision was made and the families were allowed to join them. As Turkey was not experiencing an economic upturn, a large proportion of Turkish women postponed their return and left the decision to remain in Germany open. As a result, they themselves lacked the incentive to integrate and retained a provisional identity status that became increasingly entrenched. At the beginning of the 1990s, an influx of asylum seekers caused the number of foreign women in Germany to rise within a very short space of time and to an unprecedented extent. This led to insecurity among the majority of the German population and the new arrivals felt this insecurity. Being separated from their close family ties in Turkey and living in a foreign country whose language and habits they did not know also made them appear completely ignorant and inexperienced. As they regarded the situation as temporary, they accepted the situation. Returning after the

[5] Cf. Hunn, Karin (ed.): Nächstes Jahr kehren wir zurück, die Geschichte der türkischen Gastarbeiter in der Bundesrepublik, Göttingen 2005, p. 33 f.

[6] Cf. Schmidt, Rudolf: Die Türkinnen, die Deutschen und Europa, Ein Beitrag zur Diskussion in Deutschland, Wiesbaden 2004, p. 108 ff.

recruitment stop was often perceived as too great a risk, as it meant that the path to returning to Germany was closed. After spouse and family reunification was authorised, family members also joined them. Since 1999, there has been a right to naturalisation and the children of foreign women who have lived in Germany for a long time can also acquire German citizenship. Only since this regulation has the number of naturalisations in Germany risen sharply. However, of the 1,912,169 Turkish citizens living in Germany in 2002, only around a quarter had German citizenship. Today, the figure is likely to be around 700,000. However, this figure does not include those born in Germany.
Children of parents of Turkish origin who live permanently in Germany are not included in the statistics. Despite all attempts to limit immigration, the Turkish community in Germany continued to grow steadily.[7]

Many children of Turkish origin grew up in a completely Turkish-speaking environment and only came into contact with the German language at primary school. As a result, there were primary schools, especially in larger cities, where the majority of children did not speak the language of instruction. For many Germans, this was a reason to leave their neighbourhood. As a result, the tendency to remain among themselves increased in these neighbourhoods.[8]

In the centre of Kreuzberg, around 72.6 percent of residents have a migration background.[9] In the Kottbusser Tor redevelopment area, more than ten per cent of residents north of Kottbusser Tor are Turkish nationals, while south of the square the proportion of Turkish women is

---

[7] Cf. ibid.
[8] Cf. ibid. p. 115.
[9] Cf. Senate Department for Urban Development and the Environment (ed.): Zentrum Kreuzberg/Oranienstarße, http://www.stadtentwicklung.berlin.de/wohnen/quartiersmanagement/de/nkz/index. shtml, last accessed 21.08.2014.

between seven and ten per cent.[10]

The decades-long refusal of political decision-makers to recognise immigration as such and to take the appropriate measures has meant that migrant women are often not seen by their fellow citizens as an equal part of society. This has an impact on the sense of belonging and identity of immigrant women on the one hand and on the attitude of locals towards people of other cultures and religions on the other. It is long overdue for Germans and Turks to openly discuss how they live together, as the fundamental recognition of migrants as an integral part of German society is the basis for a process of mutual understanding. This is the only way to promote conflict-free coexistence and strengthen the "we-feeling".[11]

2.2 Excursus: The oriental city

The empirical part of the thesis will compare the structure of the oriental city with the urban structures in the planning area.

In this theoretical introduction, structures of the oriental city relevant to the work are listed. The oriental city has undergone significant changes in the past. The most important differences concern the oldest, the autochthonous oriental city, the allochthonous oriental city, which was shaped by foreign influences, and the Islamic-oriental city, which was shaped by the influences of Islam. In older cities, several phases of these developments can often be seen.[12]

---

[10] Cf. Senate Department for Urban Development and the Environment (ed.): Geodata Catalogue / Soldner Berlin Y: 18586.11 X: 27215.53, Foreigners 2008 (LOR) - Turkish citizens, http://fbinter.stadt-berlin.de/fb/index.jsp, last accessed 27.08.2014.

[11] See Zimmermann, Olaf / Geißler, Theo (eds.): Kulturelle Vielfalt leben: Opportunities and challenges of intercultural education, Berlin 2012, p. 99.

[12] Cf. Kortmann-Niemitz, Ines: Merkmale derorientalischen Stadtam Beispiel der Türkei, Berlin 1990, p. 3.

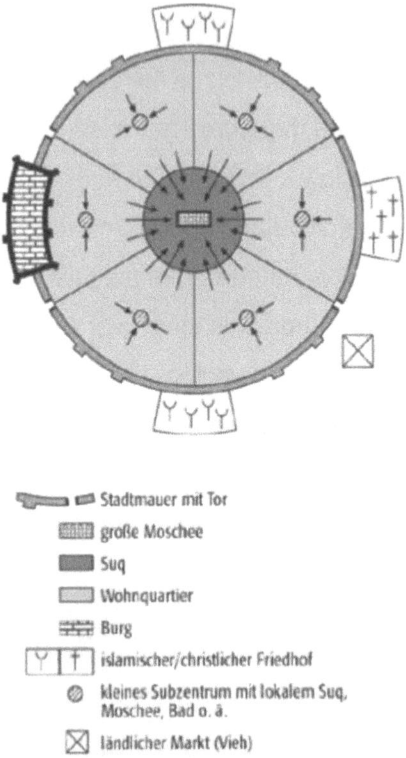

Fig. 2: Oriental city

The economic centre of Turkey's traditional oriental city is home to trade and crafts and is centrally located within residential areas. A typical feature of this so-called bazaar is its sectoral organisation. Similar businesses and shops that aim to appeal to a specific group of shoppers are located close together. This makes it easier for shoppers to get an overview, allows better organisation and coordination for sellers, facilitates competition monitoring and tax collection by the state.

There is hardly any space for storage in the shops of the bazaar, so special buildings, the hane, serve as a stacking and transhipment centre for wholesalers. They can be found in the peripheral areas of the bazaar. In addition to the main bazaar, there are also several neighbourhood bazaars, which are not used for upmarket goods but for everyday needs. The main alleys of a bazaar can be recognised in the city plan by the fact that they are wider than the residential streets and usually have a straight or only slightly curved street layout. In the oriental city, the residential neighbourhoods are divided according to the close cohesion of families, groups of the same religion, people of the same origin or people who belong together politically or economically. This spatial segregation often led to the creation of numerous neighbourhoods, which in earlier times had partial self-government.

Within a neighbourhood, sub-centres often developed with a mosque, church or synagogue, shops, public facilities and a fountain. The richer or more important inhabitants of the neighbourhood often lived near the mosque, which indicates that the living space was divided up according to social aspects within the residential areas. The cul-de-sac layout is typical of these neighbourhoods. In other words, these are narrow streets designed for pedestrian and pedal traffic, which frequently change direction and width and often end in cul-de-sacs.[13]

2.3 Cultural heritage

Cultural heritage is a much-discussed topic today and therefore the subject of international cultural policy. UNESCO has set itself the task of preserving the cultural heritage of mankind. This approach to the past takes many forms, such as monuments, landscapes, artefacts, languages, knowledge, performing arts and music. The origins of this heritage are just as diverse. People who trace their own traditional roots and acknowledge the many different influences that have shaped their own history and cultural identity are more capable of building peaceful

---

[13] VgL ibid. p. 6 ff.

relationships with other people, continuing the old dialogue and thereby forging their own future.[14] But how do traditions, spaces or things become cultural heritage? The following sections are intended to provide an insight into the different types of cultural heritage in order to initiate the subsequent discussion on the communication of cultural heritage.

Since UNESCO adopted the 'Convention concerning the Protection of the World Cultural Heritage' in 1972 and 191 countries have already signed this convention, the following definitions are based on UNESCO's understanding of cultural heritage. This international organisation is one of the most important instruments for the protection of cultural heritage.[15]

The resulting conflict potential and significance for urban planning are then discussed.

2.3.1 Material cultural heritage

Material cultural heritage is initially understood as a unique and authentic expression of a defined material heritage.[16] It could be described as an object, group or part of the past whose research and preservation are in the public interest because of its historical, scientific, artistic, urban planning or cultural landscape characterising value.[17]

The German UNESCO Commission adopted a general declaration on cultural diversity in Paris in 2001. Among other things, the UNESCO General Conference states that culture is at the centre of current debates on identity, social cohesion and the economic development of a

---

[14]Cf. Hemme, Dorothee/Tauschek, Markus/Bendix, Regina (eds.): Prädikat "Heritage", Wertschöpfungen aus kulturellen Ressourcen, Berlin 2007, p. 53.

[15] Cf. German Unesco Commission e.V. (ed.): UNESCO World Heritage Programme, http://www.unesco.de/welterbe.html, last accessed on 14.08.2014.

[16] Cf. Knoche, Michael (ed.)/Ulbricht, Justus H. (ed.), Weber, Jürgen (ed.): Das "deutsche Buch" in der Debatte um nationale Identität und kulturelles Erbe, Göttingen 2006, p. 53.

[17] See Tauschek, Markus: Kulturerbe, Eine Einführung, Berlin 2013, p.27 ff.

knowledge-based society and that respect for the diversity of cultures, tolerance, dialogue and cooperation through trust and understanding lead to peace and security. The aim is to develop cultural exchange.[18]

The protection of these cultural assets means that our relationship to them can change, allowing them to take on new functions and new meanings. It is therefore a process that can be viewed from both a historical and a contemporary perspective. It also raises the question of how cultural values are staged and how the meaning changes as a result. It is clear that this is a far-reaching concept that affects many areas of our everyday lives.[19] "Cultural heritage is based on and generates values at the same time."[20]

2.3.2 Intangible cultural heritage

For the sake of completeness, a definition of intangible cultural heritage is also provided here. However, the work will later focus on tangible heritage.

In 1982, the material understanding of what needed to be protected was expanded to include an immaterial understanding. It was now also necessary to recognise the unique spiritual and emotional heritage of mankind and to continue to protect the arts and crafts, rituals and knowledge.[21] Intangible cultural heritage includes cultural expressions such as dance and theatre, music, languages, craft techniques and oral traditions.[22] It is an expression of creativity and inventiveness, conveys

---

[18] Cf. German Commission for UNESCO (ed.): Universal Declaration of Human Rights cultural diversity, http://www.unesco.de/443.html, last accessed 27/08/2014.

[19] See Tauschek, Markus: Kulturerbe, Eine Einführung, Berlin 2013, p. 27 ff.

[20] ibid., p. 29.

[21] Cf. Knoche, Michael (ed.)/Ulbricht, Justus H. (ed.), Weber, Jürgen (ed.): Das "deutsche Buch" in der Debatte um nationale Identität und kulturelles Erbe, Göttingen 2006, p. 54.

[22] Cf. German Unesco Commission (ed.): Intangible Cultural Heritage, http://www.unesco.de/kulturerbe.html, last accessed on 13 August 2014.

identity and continuity and is based on human knowledge and expertise. As it is passed on from generation to generation, it can always be changed or remodelled.

To ensure that this traditional knowledge is preserved, UNESCO adopted the Convention for the Safeguarding of the Intangible Cultural Heritage in 2003. Just over 150 countries have now signed up to the convention.[23]

Intangible cultural heritage has no clearly identifiable originator. Heritage is created and utilised collectively and thrives on its further development, use and appropriation through the exchange processes of various actors.[24]

Popular culture is becoming the focus of politics and traditions need to be protected from 'cheap commercialisation and superficial folklore'. This also means that cultural heritage can take on new functions and thus become an important political and economic resource. National lists or inventories can also be subject to state or bureaucratic control.

The transformation into an intangible cultural heritage worthy of protection requires a conscious, reflexive act of value creation. History and historical culture or tradition and traditional culture must be considered together.[25]

2.3.3 Conflicts

The topic of cultural heritage is also a conflict-laden process. For example, dealing with historical cultural heritage raises the question of how we deal with our history. Historical culture is subject to constant change and is therefore associated with possible shifts in the meaning of

[23] Cf. German Unesco Commission (ed.): Intangible Cultural Heritage, http://www.unesco.de/immaterielles-kulturerbe.html, last accessed on 13 August 2014.
[24] See Tauschek, Markus: Kulturerbe, Eine Einführung, Berlin 2013, p. 142.
[25] See Tauschek, Markus: Kulturerbe, Eine Einführung, Berlin 2013, p. 23 ff.

the inherited heritage over time. This is also reflected in how we deal with the materiality of our cultural heritage. Materiality can conceal the 'real[1]cultural heritage'. This is particularly evident in the reconstruction after the Second World War. The destroyed components or even entire buildings can be rebuilt true to the original.

Diverging interests can clash, for example when different functions of a building are interpreted differently by different actors. The passing on of cultural assets to future generations can be jeopardised by the threat of war and destruction. Appointment procedures are also highly controversial. The ownership statuses are often very unclear: Who decides what content? Who determines what is protected or reconstructed? Who determines priorities? Which cultures are represented? Interpretative sovereignty and scope for decision-making are still barely defined and favour this dispute.

In the area of intangible cultural heritage, 'cheap commercialisation and superficial folklore' can distort or destroy traditions. However, intangible cultural heritage also compensates for the unfair distribution of tangible cultural assets perceived by the population, as it does not primarily take Western perspectives on cultural heritage into account.

On the one hand, the population's lack of knowledge, understanding and engagement with the issue leads to conflicts. On the other hand, problems can also arise due to overuse, for example through tourism. As a tangible label, the UNESCO seal guarantees cultural quality and excellence and thus opens up opportunities for tourism.

The critical combination of all the different fields, scientific perspectives, players and institutions in this context is intended to be exemplary and not exhaustive.[26]

---

[26]Cf. ibid. p. 19 ff.

2.3.4 Significance for urban planning

Cultural heritage and its communication play an important role in urban planning, as the recognition of cultural heritage represents the population and thus creates identity. The city is a historical organism, as every development has left its mark in terms of architectural history. Particularly in cities characterised by post-war modernism, the preserved heritage remains a historically defining element, even if it is only a fingerprint of the urban layout. The historic city is just as essential as written records or living traditions. Whether we are talking about cultural heritage or the "normal" historic city, structural development is always necessary and this cannot be achieved without interfering with the archive of the urban soil. As the destruction of historic buildings is irreversible, it should be checked beforehand whether planning is justifiable or unavoidable. It is the task of monument preservation and urban planning to clarify in an integrative manner which stakeholders should be involved in solving the problem and how the preservation of the cultural heritage can be reconciled with a possible change. The population must also be involved in this process, as this increases public acceptance and the success of a measure, and economisation should also be counteracted so as not to disrupt the passing on of the cultural assets to be preserved. Cultural heritage mediation breaks down linear self-perceptions and can be used to fight against assimilation efforts. In addition, historical connections between two cities can be shown through urban comparisons. This creates more intensive relationships between people and cities. They characterise the cityscape and thus reach the public.[27]

---

[27] Cf. Federal Ministry of Transport, Building and Urban Development (ed.): Stadtarchäologie und Stadtentwicklung im Welterbe, http://www.bbsr.bund.de/BBSR/DE/Veroeffentlichungen/BMVBS/Sonderveroeffentlichungen/2013/DL_Stadtarchaeologie.pdf?____ blob= publicationFile&v=2, last

Accessed on 27 August 2014, p. 25 ff.

CHAPTER 3

Kottbusser Tor redevelopment area

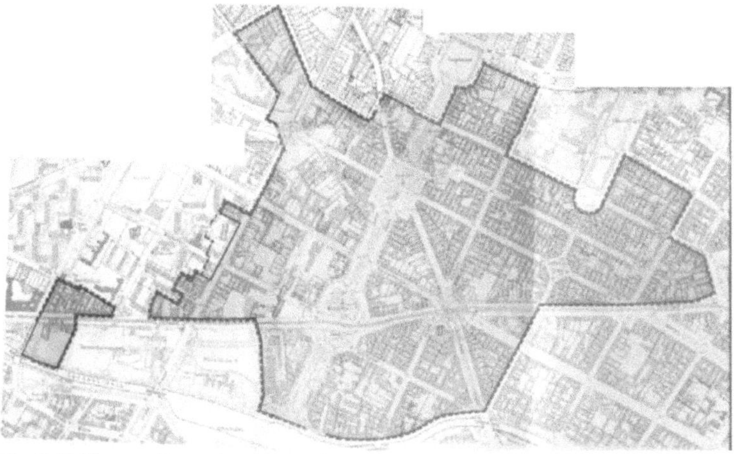

Fig. 3: Kottbusser Tor redevelopment area

After the fall of the Berlin Wall and the unification of the two halves of the city, Berlin underwent a profound transformation. Political and social structural changes shook the urban fabric. This led to contrasting developments of upheaval and decline. This was reflected in the socio-spatial structure of the city and the united Berlin developed into a city with strong social fractures. Urban development policy has confronted this situation and introduced strategies at various levels to counteract further demixing of the population and segregation processes.[28]

The redevelopment of the area around Kottbusser Tor in Kreuzberg was decided by the Senate at as part of the first urban renewal programme in 1963. It was the largest of a total of ten redevelopment areas in Berlin.

---

[28] Cf. Beer, Ingeborg / Musch, Reinfried (ed.): Berlin-Kreuzberg - Kottbusser Tor, http://www.staedtebaufoerderung.info/StBauF/SharedDocs/Publikationen/StBauF/SozialeStadt/0161_zwischenbilanz.pdf? ____blob= publicationFile&v=1, last accessed on 27 August 2014, p. 1 f.

On 4 July 1972, the redevelopment area was defined within the first legal ordinance in accordance with Section 5 of the Urban Development Act.

3.1 Refurbishment history

Until the 1980s, different models and procedures for the redevelopment of Kottbusser Tor largely coexisted. The various phases since the early 1960s can be distinguished by their different objectives and approaches.

3.1.1 First phase: area redevelopment 1965 -1975

From the mid-1960s, extensive concepts were developed for a fundamental reorganisation. The redevelopment area was divided into eleven planning units, in which large housing associations were commissioned as redevelopment agencies to manage the preparation and implementation of the renewal. The aim was to purchase land and buildings, free up flats and relocate tenants and tradespeople. Almost all of the houses were to be demolished in order to carry out a new development along the lines of the large housing estates on the outskirts of the city. Following the demolition of the old buildings, new street layouts were to be created and high-rise apartment blocks and other facilities were to be completely remodelled. The objectives and plans for this large-scale project were repeatedly changed and the proportion of old buildings to be preserved constantly increased. The result was a structure of almost incoherent building forms. [29]Following increased public protests and the resulting influence of public participation, new redevelopment concepts were finally developed. As a result, the perimeter block structure was preserved and old buildings were modernised or replaced by modern

[29] See Senate Department for Urban Development and the Environment (ed.): 23rd Report on Urban Renewal, 01.01.2000 - 31.12.2001, http://www.stadtentwicklung.berlin.de/staedtebau/foerderprogramme/stadterneuer ung/de/download/23StB-00.pdf, last accessed 23 August 2014, p. 53.

buildings. Although the cityscape was fundamentally preserved, the de-renting, conversion and displacement of long-established residents and the destruction of established areas took its course.[30]

3.1.2 Second phase: Block gutting 1972 -1979

After the "Rund um Bethanien" competition was organised in 1973, the urban planning guidelines changed.[31] The aim was now to preserve and modernise the front buildings along the streets. Following changes to the plans, the inner block development was gutted by demolishing the side wings, rear buildings and commercial buildings in favour of more generous open spaces. The social impact can be compared to a clear-cut redevelopment. After this brief transitional phase, the preservation of existing old buildings through renovation became the main task of urban renewal. On behalf of the IBA 87, urban planners and architects developed "preserving block concepts" for the redevelopment area. These included the extensive preservation of the old building fabric, including the small commercial buildings. The aim of the renovation was to integrate social and cultural infrastructure with residential, commercial and green spaces into a densely built-up residential area without destroying the historic building fabric .[32]

3.1.3 Third phase: Cautious urban regeneration from 1980

After the IBA was commissioned by the Senate, the aim was to halt the desolation of city centres in order to reclaim them as places to live. The

---

[30] Cf. Senate Department for Building and Housing (ed.): Der Block 103 in Berlin-Kreuzberg, Sanierungsgebiet Kottbusser Tor, Berlin 1994, p. 9 ff.

[31] Cf. The Senator for Building and Housing (ed.): Urban renewal around Bethanien in the Kreuzberg - Kottbusser Tor redevelopment area, Berlin 1975, p. 3

[32] VgL Senate Department for Urban Development and the Environment (ed.): 23rd Report on Urban Renewal, 01.01.2000 - 31.12.2001, http://www.stadtentwicklung.berlin.de/staedtebau/foerderprogramme/stadterneuerung/de/download/23StB-00.pdf, last accessed 23 August 2014, p. 53.

aim was to save the "broken city"[33] and the concept of "careful urban regeneration" was developed[34] . The IBA developed twelve principles for urban regeneration. These were adopted by the Kreuzberg district council in 1982. The new concept centred on more intensive resident participation and a redevelopment process that was socially acceptable. Renovations were now coordinated house by house and the 'Kottbusser Tor Renovation Commission' was set up to discuss all renovation projects with the authorities, the organisations responsible and those affected. A phased refurbishment to preserve the building fabric ensured that the affordable rents and the Kreuzberg mix of residential and commercial buildings were maintained. From today's perspective, the gradual implementation seems time-consuming and not always efficient. However, the main goal of stopping the deterioration of the housing and stabilising the controversial renewal process of the sub-area was achieved.[35]

3.2 The Kreuzberg mix

The following sections on the Kreuzberg mix and the integrated approach are presented retrospectively on the basis of the IBA 87 approach.

The Kreuzberg mix is a unique phenomenon. In this mix of housing, crafts, industry, commerce and culture, the architectural component is just as distinctive as the social mix. Immigrating population groups have repeatedly ensured that the resident structure of this traditional Berlin neighbourhood has remained in flux.[36] With the "cautious urban

[33] Senate Department for Building and Housing (ed.): Der Block 103 in Berlin-Kreuzberg, Sanierungsgebiet Kottbusser Tor, Berlin 1994, p. 10.
[34] ibid.
[35] Cf. ibid.
[36] VgL Internationale Bauausstellung Berlin (ed.): Kreuzberger Mischung, die innerstädtische Verflechtung von Architektur, Kultur und Gewerbe, Berlin 1984, p.9.

renewal"[37] , Turkish residents were also given the right to stay in the neighbourhood. This resulted in the diverse population mix that lives close together today.[38]

The residents of Kreuzberg in the 1970s were cautiously characterised schematically by means of three different groups:

The first group describes the long-established residents. These include the established residents, people who grew up here and will probably die here. They live from the neighbourhood's past, which is why their identification tends to be backward-looking.

The second group is made up of Turkish residents who were primarily looking for affordable housing when they immigrated. They do not look back or into the future. What counts for them is the present. Nevertheless, the Turkish women have developed a functioning subculture with their own meeting places, information centres and aid organisations.

The third group is more future-orientated. Although it is dependent on affordable housing in the present, it is striving for a degree of social renewal. For them, the residential value of the neighbourhood lies in the spatial, economic and social freedom. This may be the place to try out the lifestyles of tomorrow.[39]

Every group in the area was affected by unemployment and displacement tendencies due to modernisation measures. However, the Turkish residents were hit even harder, as they had fewer opportunities to react. Young people of Turkish origin usually had poorer school qualifications

---

[37] Senate Department for Building and Housing (ed.): Der Block 103 in Berlin-Kreuzberg, Sanierungsgebiet Kottbusser Tor, Berlin 1994, p. 10.
[38] Cf. ibid. p. 12.
[39] See Senator for Building and Housing (ed.): Strategies for Kreuzberg. Project description, information on the project area, objectives, Berlin 1977, p. 14 f.

than young people of German origin, which put them at a disadvantage when looking for apprenticeships and jobs. The reasons for this lay not only in language difficulties, but also in their cultural ambivalence between their home country and host country. Many Turkish women quickly became interested in self-employment. They were often dependent on German intermediaries. A lack of material resources and a lack of specialised knowledge meant that such attempts repeatedly failed. In this predicament, one's own social network is of the utmost importance.[40] These problems from the 70s and 80s still characterise the current situation in Kreuzberg, but are no longer relevant.

3.3 The integrated approach

The core idea of "cautious urban development"[41] is the integration of social and structural objectives. As a result, procedures and principles for the realisation of these objectives were developed.

3.3.1 Resident participation

Intensive participation is required to carry out planning with and for the people. For this reason, the planning and the process were coordinated with the residents. The type and scope of the modernisation measures were discussed in individual meetings and tailored to the needs of the residents as far as possible. What was particularly new from 1982 onwards was that STATTBAU, an alternative redevelopment agency that only centrally managed a small number of flats in the area, was a "resident-led"[42] redevelopment agency. This meant that the residents of

---

[40] Cf. Internationale Bauausstellung Berlin (ed.): Kreuzberger Mischung, die innerstädtische Verflechtung von Architektur, Kultur und Gewerbe, Berlin 1984, p. 274.

[41] Senate Department for Building and Housing (ed.): Der Block 103 in Berlin-Kreuzberg, Sanierungsgebiet Kottbusser Tor, Berlin 1994, p. 10.

[42] Senate Department for Building and Housing (ed.): Der Block 103 in Berlin-Kreuzberg,

a building were allowed to have a significant say in the business policy of their building.[43]

3.3.2 Rights of disposal

The owner's rights severely restrict the tenants' ability to determine their own living arrangements. This is because the homeowner decides on the floor plan, design and neighbours. When residents leave the passive status of tenant, they receive some of the rights and obligations of a private owner. A way was therefore to be found to give the residents rights and responsibility for their homes without marginalising the socially weaker . This is how the residents' cooperative 'Luisenstadt eG Grundstücksverwaltungs-genossenschaft[1]was founded.[44]

3.3.3 Other life forms

Protests against Berlin's reorganisation policy were expressed in the form of squats. The squatters' movement also tried to gain space for new forms of living and housing. Forms of flat-sharing or house-sharing emerged. Alternatives to conventional housing and living were trialled. STATTBAU adapted to the new needs of the residents and a large communal kitchen for around 60 people was built at Manteuffelstraße 40/41, for example. Flats were gradually combined to create space for the new shared flats. Roof terraces were opened up for all to use. Green rear courtyards were designed to invite people to meet and linger. Today, living in large groups has lost its appeal and partitions are once again being erected to allow people to live within their own four walls.

The squatters' movements, the building forms and the emergence of co-

Sanierungsgebiet Kottbusser Tor, Berlin 1994, p. 16.

[43] VgL ibid. p. 15 ff.

[44] Cf. ibid. p. 15 ff.

operatives have shaped the Kottbusser Tor redevelopment area and created almost village-like social spaces there.[45]

3.3.4 Integrated community facilities

A block near Heinrichplatz integrated commercial and communal facilities into the residential areas. This also resulted in many semi-public tenant gardens in the rear courtyards. Small shops and cafés were created in the front buildings. Most of them belong to the residents. Cultural projects and small businesses continue to develop in the rear courtyards. There are daycare centres and children's shops in other cooperative buildings. Green spaces and adventure playgrounds for children and adults are being created on vacant lots or brownfield sites. Neighbouring daycare centres and residents take care of the green spaces; the public sector does not provide any maintenance. The imaginative use of the narrow development and the small-scale integration of social infrastructure, green spaces and commerce not only have enormous social benefits, but also ecological ones by preventing long journeys and traffic.[46]

[45] Cf. ibid. p. 19 f.
[46] Cf. ibid. p. 19 f.

CHAPTER 4

Previous planning approaches

In the following section, relevant topics from planning approaches that have already been implemented for the stabilisation and upgrading of disadvantaged districts will be identified and presented.

4.1 Social city

The "Socially Integrative City" describes a funding programme of the Senate Department for Urban Development and the Environment and includes strategies for vibrant neighbourhoods. The aim is to qualify the city as a framework for a solidary society. Partnership processes between all levels of administration, private sector players, associations, institutions, initiatives and residents are intended to activate development potential in the neighbourhoods. Residents are to be encouraged to take responsibility for their city in order to enable the development of new perspectives. In these areas, a neighbourhood management is usually set up to take on central administrative tasks.[47]

One area designated by the 'Socially Integrative City' programme is the Kreuzberg/Oranienstraße centre. This is a large part of the redevelopment area at Kottbusser Tor. There is a primary school in the area where 99 per cent of the pupils have a migration background. The population of the area consists of 72.6 per cent residents with a migration background. The planning area is also characterised by social housing.

---

[47] Cf. Senate Department for Urban Development, Förderprogramme und -Strategien für lebendige Quartiere, http://www.stadtentwicklung.berlin.de/wohnen/quartiersmanagement/, last accessed 29.08.2014.

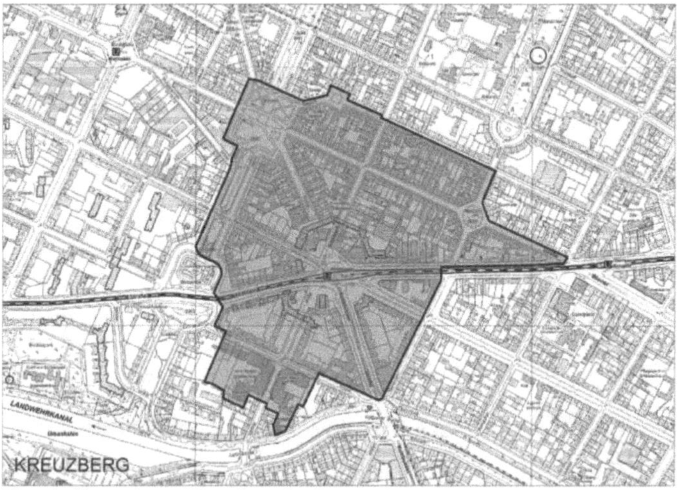

Fig. 4: Kreuzberg/Oranienstraße centre

In order to improve the educational opportunities of children with a migration background, projects were approved to promote language and speech skills. Furthermore, leisure activities for children and young people were expanded to offer meaningful leisure activities, often with educational support. The promotion of art and culture through targeted support for activities has a significant impact on the participation of residents. Projects aimed at preventing violence and avoiding conflict ensure that contact is made with teachers and promote problem-solving among migrant women. Such measures can bring the goal of integration closer. Targeted integrative activities bring people together and improve the coexistence of different cultures.[48]

4.2 Integrated urban district development concept

The 'Socially Integrative City' programme gives rise to specific integrated action plans. These form the basis for the implementation and ongoing

[48] See Senate Department for Urban Development (ed.): Zentrum Kreuzberg/Oranienstarße, http://www.stadtentwicklung.berlin.de/wohnen/quartiersmanagement/de/nkz/index. shtml, last accessed 29/08/2014.

continuation of the programmes. In addition to the locally limited and specifically project-related concepts of the 'Socially Integrative City' programme, the integrated district development concept (INSEK) is designed as a large-scale development strategy for an urban area. These strategies go beyond the individual neighbourhood management areas and focus on comprehensive socio-economic and socio-cultural problems in the city. The aim is to develop an area that takes into account not only the built space but also its social functions.[49]

In recent years, an INSEK 'Aktionsraum plus Kreuzberg-Nordosf' has been drawn up for the northern part of Kreuzberg. In spring 2010, five 'action areas plus[1]were defined by a Senate resolution. Such an 'action area plus[(1)]'contains the following general development goals:

- Improving housing and living conditions,
- Coping with the urban development consequences of demographic and economic structural changes,
- Improving educational opportunities,
- Teaching additional skills to the residents,
- Building a positive image of the neighbourhood,
- Better access to the labour market,
- Improving health opportunities

The description of the main fields of action is based on four interdisciplinary working groups (FAG):

- Education, youth and training (FAG 1),

[49] See Senate Department for Urban Development and the Environment (ed.): Berlin, Integrierte Stadtteilentwicklung (Soziale Stadt), Integriertes Stadtteilentwicklungskonzept für den Aktionsraum Kreuzberg - Nordost, http://www.stadtentwicklung.berlin.de/soziale_stadt/aktionsraeume_plus/download /INSEK_Kreuzberg-Nordost_Internetklein.pdf, last accessed 29 August 2014, p. 6.

- Labour and economy (FAG 2)
- Culture, community (social affairs and health) and integration (FAG 3)
- Neighbourhoods and public spaces (FAG 4)

The 'action area plus' is understood by all those involved as a learning process. 'Socially Integrative City' funds are being made available for the realisation of this process. These are intended to help establish and test cooperation and to jointly develop and realise innovative, experimental and interdisciplinary measures. The following section focuses exclusively on the guiding principles, objectives and priorities of FAG 3 Culture, Community and Integration in order to develop further ideas for approaches to communicating Turkish cultural heritage.[50]

According to the INSEK Kreuzberg-Nordost, the area's strengths in terms of social space lie mainly in its heterogeneous social structure, which has a high percentage of families. However, educationally aware families with young children migrate at an above-average rate and leave Kreuzberg-Nordost before their children start school.

The INSEK developed the mission statement in 2010

'Building bridges - opening doors - strengthening networks[1].

---

[50] Cf. Senate Department for Urban Development (ed.): Integriertes Stadtteilentwicklungskonzept, Aktionsraum Kreuzberg-Nordost, May 2012, http://www.stadtentwicklung.berlin.de/soziale_stadt/aktionsraeume_plus/download ZINSEK_Aplus_Kreuzberg_2012.pdf, last accessed 10 September 2014, p. 1.

Fig. 5: Mission statement "Building bridges - opening doors - strengthening networks"

The current update focuses on the realisation of an integrated and socially sustainable development of the action area. The aim is to build bridges between families and educational institutions, different cultures, generations, long-established residents and immigrants and to create a social neighbourhood with strong partners. Open Gates[1]is a call to politicians, administrations and institutions to enable all residents to participate actively and equally in social, political and economic life. Existing networks should be expanded and, if necessary, new network structures should be established. These networks form the basis of social activities.[51]

FAG 3 Culture, Community and Integration in Kreuzberg-Northeast has set itself the goals of 'ensuring a needs-based social infrastructure[1]and

[51] Cf. ibid. p. 84 ff.

'promoting equal access opportunities to social services for all residents'. The social meeting places in the planning area include neighbourhood and district centres, multi-generational houses, family centres and youth facilities. These facilities are important places that function as a basis for social interaction and as places of participation.

A thematic and spatial linking of different offers should be supported to achieve these goals. As a prerequisite for this, it is necessary to expand, network and strengthen social meeting points for the neighbourhood. The cultural offerings should be further developed and outreach counselling and outreach should be consolidated and expanded in terms of location.[52]

Whether these approaches of the 'Socially Integrative City' or the INSEK could also represent approaches for the mediation of Turkish cultural heritage in a figurative sense will be clarified later in the work.

---

[52] VgL ibid. p. 99 f.

CHAPTER 5

Inventory of Turkish cultural heritage

In the following section, examples of architectural and spatial Turkish cultural heritage in the Kottbusser Tor redevelopment area are presented in order to later illustrate the approaches to communicating Turkish cultural heritage. Furthermore, it will be examined whether the structures of the oriental city have parallels with the urban structure of Kottbusser Tor.

5.1 Architectural and spatial cultural heritage

Celalettin Kesim Monument

Fig. 6: Celalettin Kesim memorial

Celalettin Kesim came to West Berlin from Turkey in 1973. He worked as a lathe operator at Borsig and was a shop steward for IG Metall. His last job was as a teacher at a vocational school. His pregnant wife and a child were waiting for him in Turkey.

On 5 January 1980, around 20 activists from the politically left-wing Turkish Women's Centre at Kottbusser Tor distributed leaflets against the impending military coup in Turkey. They were attacked by around 70 Turkish right-wing extremists from the Grey Wolves and religious fundamentalists. There were chase scenes at Kottbusser Tor and at the underground station. Many escaped with minor injuries. Celalettin Kesim was stabbed in the thigh. After half an hour, a fire engine took him to Urban Hospital, where he was pronounced dead.

For several years, a memorial on the corner of Reichenberger Strasse and Kottbusser Strasse has commemorated Celalettin Kesim.[53] The stone bears a quote by the Turkish poet Nazim Hikmet: "They are the enemies of hope, beloved."[54]

Kottbusser Straße 8

Fig. 7: KottbusserStr. 8

The house at Kottbusserstraße 8 was owned by the Gemeinnützige Wohnungsbau- und Siedlungsgesellschaft (GSW), according to the building registry . In March 1981, the house was illegally renovated by

[53] Cf. Deniz Yünel: For the love of Allah, http://www.trend.infopartisan.net/trd0105/t360105.html, last accessed 04/09/2014.

[54] ibid.

eight women, most of them of Turkish origin, with their five children.

This was followed by violent confrontations with GSW construction workers, and the women were threatened with racist and Nazi slogans.

Fig. 8: Sevim Celebi-Gottschlich

Eventually, an agreement is reached with GSW.[55] n 1995, Berlin's only Kurdish mosque, the Selahaddin Eyyubi Mosque, is built on the first floor of the building. This in turn leads to tensions between Turkish and Kurdish tenants and attacks by Turkish nationalists. In 2008, the tenants of the mosque were fed up with the disputes and moved the mosque to an old factory building at Boppstraße 10 in Kreuzberg. One of the squatters, Sevim Celebi- Gottschlich, is the first woman of Turkish origin to sit in a German parliament.[56]

Today, the building houses a bicycle shop on the ground floor and flats on the upper floors.

[55] See District Office Friedrichshain-Kreuzberg of Berlin, Urban Development Office, Building Inspectorate; building file archive for Kottbusser Straße 8.

[56]Cf. Rennefanz, Sabine / Berliner Zeitung (ed.): Die grüne Festung, http://www.berliner-zeitung.de/archiv/am-kottbusser-tor-schuetzt-sich-ein-haus- pre-foxes--fixers-and-germans-hit-the-green fortress, 10810590,10631220.html, last accessed 04.09.2014.

Garden of the Kreuzberg Museum

Fig. 9: Garden of the Kreuzberg Museum

In the 1970s, the open space at what is now Adalbertstraße 95a was part of the new Kreuzberg centre and was to be converted into a car park. Squatters planted maple trees there. The Turk Remzi Karatas, who also came to Kreuzberg in the 1970s, followed the events, helped to care for the trees and successfully defended them against all construction measures.

The trees are still standing today.

In 1984, the refurbishment of the front building of Adalbertstraße 95 began on behalf of the Gewerbesiedlungs-Gesellschaft (GSG). Remzi Karatas began planting the area on the gable facade facing Adalbertstraße 95a during the course of the clean-up work. Hidden behind a construction trailer, the garden initially remained undiscovered. After the building work began, the small garden was tolerated without further ado and Remzi Karatas was granted a special right of use. To this day, he plants fruit, vegetables, herbs and flowers in his garden. There is a memorial in the garden. In 1972, the sculptor Volker Dierkes saved the original double door of the apartment block at Naunynstraße 11a from destruction. Dierke remodelled the door and transferred a photo of the façade onto a stone. The idea was to erect a gravestone for the house.[57] This monument, which Dierkes donated to the museum, was inaugurated in 1999 and is a reminder of the times of "cautious urban renewal".[58]

Aziz Nesin Primary School

[57] Cf. Brückenbauer, Friedrichshain-Kreuzberg interkulturell (ed.): Ich erhielt ein Sondernutzungsrecht May/June 2012, http://www.sekis-berlin.de/fileadmin/files/selko/downloads/Kontaktstellen/Brueckenbauer_2012- 03.pdf, last accessed 04.09.2014.

[58] Senate Department for Building and Housing (ed.): Der Block 103 in Berlin-Kreuzberg, Sanierungsgebiet Kottbusser Tor, Berlin 1994, p. 10.

Fig. 10: Ali Aziz Efendi memorial plaque

The Aziz Nesin Primary School is one of Berlin's state European schools (SESB) with the language combination German-Turkish. It was opened in 1995. Around 400 pupils attend the school. German and Turkish are taught here as equal partner languages. The aim is intercultural learning. Children with different mother tongues are taught together. In 2001, the school was named after an important Turkish writer, Aziz Nesin.[59] The school is located on the site of the first Turkish cemetery in Berlin. In the front garden of the school there is a memorial plaque set in stone to Ali Aziz Efendi, the son of Aziz Nesin, an Ottoman ambassador to Prussia and writer who died in 1798, who was regarded as a pioneer of Turkish modernism.[60]

4.3 Excursus: Structure of the oriental city

This chapter aims to identify possible parallels between the structure of the oriental city and the structure of the Kottbusser Tor redevelopment area. As there is no academic literature on this comparison, it will begin with our own empirical approaches.

It is hardly possible to define a generally valid European urban structure in the 21st century. The Kottbusser Tor redevelopment area, on the other hand, is characterised by Wilhelminian-style block development. In contrast to the oriental city, the economic centre does not exist solely in the centre of the city, because in Kreuzberg housing and manufacturing industry are closely intertwined. Nevertheless, due to its central location within Berlin, the mixed-use area is a self-sufficient economic centre in its

[59] Cf. Aziz Nesin Primary School: Concept of the school, http://www.aziz-nesin-g.cidsnet.de/konzept/konzept_e.html, last accessed 04/09/2014.
[60] Cf. Landespressedienst (ed.): Memorial plaque for Ali Aziz Efendi, http://www.gedenktafeln-in-berlin.de/nc/gedenktafeln/gedenktafel-anzeige/tid/ali- aziz-efendi/, last accessed 04/09/2014.

own right.[61] In contrast to the oriental city, businesses and shops are colourfully mixed and not organised according to sector. In addition to the main shopping streets in the redevelopment area, there are also smaller streets, like the bazaars in the oriental city, in which the shops can cover the daily basic needs of the residents. In the European city, this is probably due to the urban planning model of the "city of short distances". As in the oriental city, the main streets have a wider and mostly straight layout, whereas the side streets can also have a more curved layout.

In the oriental city, the neighbourhoods are structured on the basis of a close cohesion of people. This spatial segregation leads to the creation of numerous neighbourhoods. In the Kottbusser Tor redevelopment area, this segregation is due to social segregation. This means that the socio-spatial unequal distribution is caused by socio-economic characteristics, which can happen voluntarily or involuntarily.[62]

Within the neighbourhoods, in the oriental city and in the redevelopment area, sub-centres often emerge with their own public facilities, shops or meeting places. In these sub-centres, the more important residents often live near mosques. In Kreuzberg, the mosques are replaced by green spaces, squares or other meeting places. Here, too, the socially better-off residents live closer to the open spaces than the lower classes. In this case, both structures indicate that housing is organised along social lines.

Despite the very different urban structures of the two models, there are a few overlaps. However, these overlaps should not be interpreted as cultural heritage, as the Gründerzeit structure around Kottbusser Tor

---

[61] See Senator for Building and Housing (ed.): Strategies for Kreuzberg. Project description, information on the project area, objectives, Berlin 1977, p. 14.

[62] Cf. Farwick, Andreas: Segregation und Eingliederung, Zum Einfluss der räumlichen Konzentration von Zuwanderern auf den Eingliederungsprozess, Wiesbaden 2009, p. 383.

already existed before the wave of immigration of the Turkish population and was therefore not influenced or even shaped by the Turks. The few parallels between the two structures can only provide a good basis for facilitating dialogue and exchange between two cultures with partly similar urban development patterns. This is because the spatial structure of the cities is a reflection of the respective way of life of the population. A change in the way of life, in relation to the desired appropriation of knowledge about the Turkish cultural heritage, could therefore also trigger a change in the urban planning structures. The collective appropriation of each other's cultural heritage would then lead to a series of commonalities with regard to the character of the Kottbusser Tor redevelopment area and the quality of the cityscape, and possibly give rise to a kind of 'shared cultural heritage' in the future.

CHAPTER 6

Approaches to communicating Turkish cultural heritage

In the final part of the thesis, conceptual approaches to the mediation of Turkish cultural heritage at Kottbusser Tor will be developed on the basis of the theoretical and empirical foundations. After a description of the necessity of cultural heritage mediation, perspectives on the mediation of cultural heritage by means of participatory procedures are discussed. In addition, individual key objectives are formulated, which are based on the planning approach of the integrated neighbourhood development concept.

6.1 Necessity

"Diversity is wealth"[63] is a well-known slogan in the UNESCO context. Ethnic and cultural diversity is created by engaging with cultural heritage. This raises public awareness of the importance of cultural heritage and promotes social cohesion between different cultures.[64] It is about recognising, managing and preserving material assets, monuments and buildings that reflect a past. Shared values and collective memories are preserved. In order to preserve the sense of accumulated and shared experience and thus promote integration, cultural heritage must be preserved and passed on.[65]

Ultimately, the aim is to achieve a paradigm shift from the static model of national culture to a networked model that transcends political, national

---

[63] Zimmermann, Olaf / Geißler, Theo (eds.): Kulturelle Vielfalt leben: Opportunities and challenges of intercultural education, Berlin 2012, p. 67.

[64] Cf. ibid.

[65] Cf. Hemme, Dorothee/Tauschek, Markus / Bendix, Regina (eds.): Prädikat "Heritage", Wertschöpfungen aus kulturellen Ressourcen, Berlin 2007, p. 338.

and ideological boundaries. It is necessary to strive for changes in the communication of individuals, i.e. society and its social context should be made transparent for other cultures.

Particularly in the Kottbusser Tor redevelopment area, people talk about Germans on the one hand and Turks on the other. This ignores the fact that every person is part of cultural development and is also characterised and influenced by their cultural environment. This encounter refers not only to two cultures meeting, but rather to every interaction between individuals and groups in our society.[66]

6.2 Cultural heritage mediation and participation

The term participation stands for the active involvement, participation and inclusion of population groups, organisations and associations in decisions that affect their lives. It contributes to the target groups being able to articulate and assert their interests in development cooperation. People are given the opportunity to contribute their experiences and values to the joint work.[67]

The inclusion of Turkish citizens in the questions and solution scenarios of cultural heritage mediation work is now a matter of course in cultural work; particularly in the redevelopment area at Kottbusser Tor, it is an important part of 's work. As early as 1977, the IBA's urban development concept in the 'Strategies for Kreuzberg' competition was orientated towards the goal of promoting the coexistence of the different population groups in the area through the active participation of residents. The integrated approach of "cautious urban renewal"[68] enabled identification

[66] See Zimmermann, Olaf / Geißler, Theo (eds.): Kulturelle Vielfalt leben: Opportunities and challenges of intercultural education, Berlin 2012, p. 74.

[67] Cf. Federal Ministry for Economic Cooperation and Development (ed.): Partizipation, http://www.bmz.de/de/service/glossar/P/partizipation.html, last accessed 16 September 2014.

[68] Senate Department for Building and Housing (ed.): Der Block 103 in Berlin-Kreuzberg,

with the planning area.

To this day, addressing and integrating immigrant women from Turkey is an important aspect of urban planning and cultural work. Residents' representatives and tenant counselling sessions, which are also held in Turkish, contribute to this.

In the Kottbusser Tor redevelopment area, for example, a Turkish-German cultural centre was opened in the courtyard of Oranienstraße 34, which offers citizens a comprehensive programme of support and advice. Through youth and cultural centres in the neighbourhood, young migrants and German youths can work together on current projects by means of extracurricular educational programmes. This promotion of intercultural youth work creates a sense of togetherness at an early stage and at the same time promotes mutual understanding of the other culture. The implementation of participation processes such as neighbourhood conferences or future workshops can also ensure targeted participation.[69]

If the urban public is involved in an ongoing project, public acceptance of a project can be increased enormously. This is why participation has become essential for the success of a cultural education programme.

Making projects visible has an impact on people's immediate living environment. In cultural heritage mediation, participation helps to recognise potential conflicts at an early stage. Misunderstandings can be cleared up quickly or avoided altogether. Concrete design suggestions for current or future projects are taken on board and the various planning requirements can be integrated into the project. The legitimisation and acceptance of a project is also improved through early participation.

Sanierungsgebiet Kottbusser Tor, Berlin 1994, p. 10.

[69] Cf. Dame, Thorsten / Herold, Stephanie / Salgo, Andreas (eds.): Re-Vision-IBA '87. 25 Jahre Internationale Bauausstellung, Berlin 1987, p. 27 f.

Intensive dialogue can create trust and understanding for the other perspective among all those involved.

Participation motivates participants to take on responsibility and secure a project in the long term. The resulting identification with a project leads to long-term civic engagement, which makes an important contribution to the development of self-supporting and self-organised structures in the neighbourhood.[70]

The aim is to create a social environment for children in the planning area, with the involvement of their parents, that promotes their development from birth to the start of their careers. Particular attention is paid here to educational transitions. By teaching cultural heritage at an early age, consideration for minorities is strengthened and different views can be respected from an early age. Co-operation with other main Senate administrations, such as the Senate Department for Education, Science and Research, can support and strengthen the content of cultural heritage education projects.[71]

6.3 Integrated neighbourhood development concept

The aim now is to develop guidelines for the mediation of the Turkish cultural heritage in the Kottbusser Tor redevelopment area using the approaches of the INSEK 'Aktionsraum plus Kreuzberg-Nordost[(1)]. The aim is to draw up a schematic development strategy for the redevelopment area.

Due to the heterogeneous social structure in the planning area and the

---

[70] Cf. Senate Department for Urban Development and the Environment (ed.): Handbuch zur Partizipation - Partizipation bietet Chancen, http://www.stadtentwicklung.berlin.de/soziale_stadt/partizipation/download/Handbuch_Partizipation.pdf, last accessed 16 September 2014, p. 59 f.

[71] Cf. Internationale Bauausstellung Berlin (ed.): Kreuzberger Mischung, die innerstädtische Verflechtung von Architektur, Kultur und Gewerbe, Berlin 1984, p. 319 f.

high percentage of families, the guiding principle of 'building bridges - opening doors - strengthening networks[1]can also be applied to the problem of communicating cultural heritage. Bridges must be built between the different cultures, in this case the different generations of Turkish and German women, the families and existing educational institutions, in order to convey the existing Turkish cultural heritage and create a social neighbourhood with integrated immigrants. In order to enable equal participation in the history and information of the cultural heritage and thus gain the interest and understanding of the population, the 'gates' of politics, administrations and institutions in the neighbourhood should also be opened here.

It is necessary to secure the basis of social activities. It is therefore important to expand existing networks in order to ensure the constant dissemination of information on cultural heritage. Equal access to social services and meeting places such as neighbourhood and district centres, multi-generation houses, family centres and youth facilities support this project. These places of participation and social interaction should be linked thematically and spatially in order to ensure the goal of comprehensive communication of Turkish cultural heritage. The cultural offerings should become social meeting places in the neighbourhood through better advice and regular contact and strengthen the communication of cultural heritage. They can, for example, help to promote the exchange of ideas through projects aimed at communicating Turkish cultural heritage. Through the participation of German and Turkish young people and adults, intercultural dialogue can grow and a better understanding of each other can be achieved. In this way, long-term commitment can be anchored and content-related goals from relevant subject areas can be focussed more strongly. In order for such projects to emerge, it is necessary to develop intercultural and social skills in order to further develop and support the educational networks. It is becoming apparent that networking, counselling and activation have a very high priority in cultural mediation work. This means that local requirements for action must be formulated and short communication channels must be created, promoted and expanded.[72]

The evaluation of projects by the senate administrations and district offices is a prerequisite for achieving sustainability of the new structures. This requires long-term security and efficiency. If this is guaranteed, a broad impact and target group can be reached.

The success of the projects will largely depend on the extent to which the structures created can be stabilised. This can be achieved by organising regular joint meetings, sessions and other events. In this way, a project will be recognised as useful more quickly and will be more popular among the population.[73]

To summarise, it is important to formulate the core messages of cultural heritage mediation and to translate and communicate them to civil society. Translating means not only communicating the core messages, but also finding starting points so that a binding effect can develop at the following levels:

Build competences

This is a cross-cutting task for society as a whole, which belongs in the

---

[72] Cf. Senate Department for Urban Development (ed.): Integriertes Statdteilentwicklungskonzept, Aktionsraum Kreuzberg-Nordost, May 2012, http://www.stadtentwicklung.berlin.de/soziale_stadt/aktionsraeume_plus/download ZINSEK_Aplus_Kreuzberg_2012.pdf, last accessed 10 September 2014, p. 84 ff.

[73] Cf. Internationale Bauausstellung Berlin (ed.): Kreuzberger Mischung, die innerstädtische Verflechtung von Architektur, Kultur und Gewerbe, Berlin 1984, p. 313 f.

areas of education, further education and training.

Sustainability

Cultural education must be sustainable to ensure continuity and quality.

Ensuring participation

Cultural participation must be ensured in order to develop curiosity about one's own. This then gives rise to curiosity about the other.
This is how the guidelines for communicating the Turkish cultural heritage in the planning area are developed. Acceptance by society as a whole depends on the self-image of cultural coexistence and living together.[74]

6.4 Placement problem

In an immigration society like the one we have in Germany, especially in Kreuzberg, new identities are constantly growing or old ones are beginning to change. This makes some conflicts visible, because it brings with it a certain degree of confusion. Different cultural values, traditions, ways of life and everyday practices are becoming increasingly differentiated. Cultural institutions have the potential to respond constructively to this confusion. They have the opportunity to highlight and reflect on diversity and change.

However, this is hardly utilised because it hardly reaches a large part of our population.[75]

If the potential is utilised, the preservation of cultural heritage also always means a selection, because not everything can be honoured and some things must be forgotten in order to increase the identification potential of

[74] See Zimmermann, Olaf / Geißler, Theo (eds.): Kulturelle Vielfalt leben: Opportunities and challenges of intercultural education, Berlin 2012, p. 75 f.

[75] See Zimmermann, Olaf / Geißler, Theo (eds.): Kulturelle Vielfalt leben: Opportunities and challenges of intercultural education, Berlin 2012, p. 309.

the chosen one. The identification potential of cultural heritage therefore harbours the greatest potential for conflict. For who makes this selection? [76]

Conflicts also arise from the interaction of different cultures that have so far only touched each other on the surface. It is problematic to simply apply the patterns developed by Germans to the lives of Turkish women in Kreuzberg. It is not enough to publish brochures or reorganisation plans with German and Turkish titles where it is clear that the content is only written in German. Educational work must take place in advance, because there is a hierarchy in our system of thought in which Germans claim the power to make decisions. Residents with a migrant background are still underrepresented in cultural institutions, so there are fewer opportunities to represent the foreign population.[77]

---

[76] Cf. Hemme, Dorothee/Tauschek, Markus / Bendix, Regina (eds.): Prädikat "Heritage", Wertschöpfungen aus kulturellen Ressourcen, Berlin 2007, p. 338.

[77] Cf. Internationale Bauausstellung Berlin (ed.): Kreuzberger Mischung, die innerstädtische Verflechtung von Architektur, Kultur und Gewerbe, Berlin 1984, p. 315 f.

CHAPTER 7

Outlook

Finally, recommendations for action and forecasts for the mediation of Turkish cultural heritage will be drawn up.

7.1 Recommendations for action

Cities are places where diverse types of cultural heritage become public. This density as well as the coexistence and interaction of different cultural players is an important potential for urban development in the Kottbusser Tor redevelopment area. It presents politicians with new challenges. Kreuzberg is characterised by local internationality and offers living spaces for people with different migration backgrounds. These aspects of cultural diversity need to be protected and promoted. Internationalisation through cultural and social interdependencies to the outside demands the same in terms of cultural openness to the inside. This includes clearly communicating to the urban public that cultural heritage is an enrichment for the area, which in turn means that cultural heritage must become the subject of all municipal policy fields of action. Although a lively intercultural project community already exists in the Kottbusser Tor redevelopment area, it is not sufficient for the sustainable protection and promotion of the new project to communicate Turkish cultural heritage. Rather, it is about structural and permanent

Changes in urban cultural heritage mediation.

Cultural institutions must find new ways to continuously and successfully adapt to social change.

This requires the development of municipal strategies. These can be developed in the form of guiding principles or guidelines and development planning in the short, medium and long term. Strategies can be developed on the basis of research in order to specifically address new user groups in the future. This will enable the facilities to open up interculturally in the future by adapting programmes or content. Regular public events can also raise public awareness of the Turkish cultural heritage. [78][79]

Furthermore, it is necessary to loosen up the total rejection of religion, for example among atheists, in order to counteract the formation of islands in the area and thus avoid a monopoly situation. It is necessary to at least address the need for religion.[80]

7.2 Forecast

In 2006, UNESCO formulated and set out the goal of "cultural education for all". This convention should act as a guideline and focus for the cultural centres in the future. Once the binding and legal protection of the programmes and facilities in the Kottbusser Tor redevelopment area has been achieved, they can promote an effective strategy for communicating Turkish cultural heritage. With this convention, all contracting parties would commit to bringing cultural heritage to the public through educational programmes. This cultural education would help to ensure that the objectives discussed in the previous chapters would be sustainable. By imparting knowledge, skills and abilities and by

---

[78] See Deutsche UNESCO-Kommission e.V. (ed.): Kulturelle Vielfalt gestalten-Handlungsempfehlungen aus der Zivilgesellschaft zur Umsetzung des UNESCO-Übereinkommens zur Vielfalt kultureller Ausdrucksformen (2005) in und durch Deutschland, http://www.unesco.de/fileadmin/medien/Dokumente/weissbuch_lay_endf_internet. pdf, last accessed 17 September 2014, p. 11 f.

[79] Cf. ibid. p. 13.

[80] Cf. Internationale Bauausstellung Berlin (ed.): Kreuzberger Mischung, die innerstädtische Verflechtung von Architektur, Kultur und Gewerbe, Berlin 1984, p. 316 f.

developing competences, attitudes and attitudes, the Turkish cultural heritage would be passed on, thus ensuring peaceful coexistence of different cultures in the neighbourhood in the long term. Once awareness is established in the population, knowledge about cultural heritage can be quickly expanded and personal and, above all, collective identities and values can be enhanced. Cultural heritage is part of lifelong learning and therefore part of general education, which should enable everyone to participate in society and play an active role in shaping it.
Cultural education and thus the dissemination of cultural heritage can only be guaranteed from the outset with lifelong commitment from civil society.[81]

Cultural diversity and the communication of cultural heritage are already on the local political agenda. If the importance and relevance of Turkish cultural heritage in the Kottbusser Tor redevelopment area were widely communicated and made visible, social conflict could be minimised. Better coexistence between the two ethnic groups would be promoted. The implementation of the conceptual solutions, which are based on the previous planning approaches of the 'Socially Integrative City' and the INSEK, would favour this exchange of culture and interests. At the same time, the identification of the Turkish population with the planning area would be promoted.

---

[81] See Deutsche UNESCO-Kommission e.V. (ed.): Kulturelle Vielfalt gestalten-Handlungsempfehlungen aus der Zivilgesellschaft zur Umsetzung des UNESCO-Übereinkommens zur Vielfalt kultureller Ausdrucksformen (2005) in und durch Deutschland, http://www.unesco.de/fileadmin/medien/Dokumente/weissbuch_lay_endf_internet. pdf, last accessed 17 September 2014, p. 27 f.

CHAPTER 8

Conclusion

The content of this thesis deals with Turkish cultural heritage and how it is conveyed to the residents of the Kottbusser Tor redevelopment area. In order to be able to correctly categorise the various dimensions of the area under investigation, the basic chapter of the thesis uses various theoretical approaches to describe the immigration of the Turkish population into the planning area, the oriental city and approaches to defining cultural heritage.

The characteristics of the Kottbusser Tor redevelopment area listed in the empirical part of the work also play an important role due to the special social perspectives of the area. Not only is a spatial reference to the neighbourhood established, but the social structures and their potential for development and change are also shown in connection with previous planning approaches.

Through the exemplary presentation of the Turkish cultural heritage in the redevelopment area, the conceptual part of the work develops an approach to communicating the Turkish cultural heritage that is based on the planning approach of the INSEK.

The result shows that the existing planning approaches, adapted and expanded, can also be applied to the mediation of cultural heritage. The concept proposed here also takes into account the results and findings of the previous parts of the work.

It is shown that the mixture of opposites in the Kottbusser Tor redevelopment area is characteristic and thus characterises the cityscape in the area. The aim is therefore to strive for mutual acceptance of each other through targeted communication of the Turkish cultural heritage. For the joint development of a project, it is important that the cultures come together from an early age in order to prevent future social tension and create communities.

If Turkish cultural heritage is to be recognised, it is important to engage with the actors who generate these processes and with their intentions so that the contradictions that separate two cultures can be bridged. In this way, cultural heritage can be integrated into everyday life. Ultimately, knowledge transfer is required. This is the only way to appropriately emphasise the cultural specificity of this locally borne cultural heritage.[82]

In this work, Turkish cultural heritage is interpreted as a resource, staged and utilised accordingly. This resource can promote positive social development through cultural diversity and enable peace and democracy.[83]

The city must be understood as a human community whose identification can be achieved through social motivation. Through the increased sharing of information and memories, eye contact between the two cultures can be constantly maintained and the other living environment can be recognised more and more.

A web of modes of appropriation and interpretation through lively historical editing transports ideologies among each other. This leads to the preservation and appreciation of the legacies of the past.

---

[82] Cf. Hemme, Dorothee / Tauschek, Markus / Bendix, Regina (eds.): Prädikat "Heritage", Wertschöpfungen aus kulturellen Ressourcen, Berlin 2007, p. 340 f.

[83] Cf. ibid. p. 187 f.

CHAPTER 9

Sources

Bibliography

Dame, Thorsten / Herold, Stephanie / Salgo, Andreas (eds.): Re-Vision-IBA '87. 25 Years of International Building Exhibition, Berlin 2011

The Senator for Building and Housing (ed.):
Urban regeneration around Bethanien in the Kreuzberg - Kottbusser Tor redevelopment area, Berlin 1975

Farwick, Andreas: Segregation and Integration, On the Influence of the Spatial Concentration of Immigrants on the
Integration process, Wiesbaden 2009

Hemme, Dorothee / Tauschek, Markus / Bendix, Regina (eds.):
Prädikat "Heritage", Wertschöpfungen aus kulturellen Ressourcen, Berlin 2007

Hunn, Karin (ed.): Nächstes Jahr kehren wir zurück, die Geschichte der türkischen Gastarbeiter in der Bundesrepublik, Göttingen 2005

Internationale Bauausstellung Berlin (ed.): Kreuzberger Mischung, die innerstädtische Verflechtung von Architektur, Kultur und Gewerbe, Berlin 1984
Knoche, Michael / Ulbricht, Justus H. / Weber, Jürgen (eds.):
The "German book" in the debate about national identity and cultural

heritage, Göttingen 2006

Kortmann-Niemitz, Ines: Merkmale der orientalischen Stadt am Beispiel der Türkei, Berlin 1990

Schmidt, Rudolf: Die Türkinnen, die Deutschen und Europa, Ein Beitrag zur Diskussion in Deutschland, Wiesbaden 2004

Senator for Building and Housing (ed.): Strategies for Kreuzberg. Project description, information on the project area, objectives, Berlin 1977

Senate Department for Building and Housing (ed.): Block 103 in Berlin-Kreuzberg, redevelopment area Kottbusser Tor, Berlin 1994

Tauschek, Markus: Cultural Heritage, An Introduction, Berlin 2013

Zimmermann, Olaf / Geißler, Theo (ed.): Kulturelle Vielfalt leben: Opportunities and challenges of intercultural education, Berlin 2012

Internet

Aziz Nesin Primary School: Concept of the school, http://www.aziz-nesin-g.cidsnet.de/konzept/konzept_e.html, last accessed 04/09/2014

Beer, Ingeborg / Musch, Reinfried (eds.): Berlin-Kreuzberg - Kottbusser Tor,
http://www.staedtebaufoerderung.info/StBauF/SharedDocs/Publik

ationen/StBau F/SozialeStadt/0161 _zwischen bilanz. pdf? blob=p ublicationFile&v=1, last accessed on 27/08/2014

Brückenbauer, Friedrichshain-Kreuzberg interkulturell (ed.): Ich erhielt ein Sondernutzungsrecht Mai/Juni 2012, http://www.sekis-berlin.de/fileadmin/files/selko/downloads/Kontaktstellen/Bruecken bauer_2012-03.pdf, last accessed 04.09.2014

Federal Ministry of Transport, Building and Urban Development (ed.): Stadtarchäologie und Stadtentwicklung im Welterbe, http://www.bbsr.bund.de/BBSR/DE/Veroeffentlichungen/BMVBS/S onderveroeffentlichungen/2013/DL_Stadtarchaeologie.pdf? _ blob =publicationFile&v=2, last accessed on 27/08/2014

Federal Ministry for Economic Cooperation and Development (ed.): Participation,

http://www.bmz.de/de/service/glossar/P/partizipation.html, last accessed 16/09/2014

Deutsche UNESCO-Kommission e.V. (ed.): Intangible Cultural Heritage, http://www.unesco.de/kulturerbe.html, last accessed on 13/08/2014

Deutsche UNESCO-Kommission e.V. (ed.): Intangible Cultural Heritage, http://www.unesco.de/immaterielles-kulturerbe.html, last accessed on 13/08/2014

Deutsche UNESCO-Kommission e.V. (ed.): UNESCO World Heritage Programme, http://www.unesco.de/welterbe.html, last accessed on 14.08.2014

Deutsche UNESCO-Kommission e.V. (ed.): Universal Declaration on Cultural Diversity, http://www.unesco.de/443.html, last accessed 27 August 2014

Deutsche UNESCO-Kommission e.V. (ed.): Kulturelle Vielfalt gestalten-Handlungsempfehlungen aus der Zivilgesellschaft zur Umsetzung des UNESCO-Übereinkommens zur Vielfalt kultureller Ausdrucksformen (2005) in und durch Deutschland, http://www.unesco.de/fileadmin/medien/Dokumente/weissbuch_la y_endf_internet.pdf, last accessed 17 September 2014.

Landespressedienst (ed.): Memorial plaque for Ali Aziz Efendi, http://www.gedenktafeln-in-berlin.de/nc/gedenktafeln/gedenktafel-anzeige/tid/ali-aziz-efendi/, last accessed 04.09.2014

Rennefanz, Sabine / Berliner Zeitung (ed.): Die grüne Festung, http://www.berliner-zeitung.de/archiv/am-kottbusser-tor-schuetzt- sich-ein-haus-vor-fuechsen--fixern-und-deutschen-schlagern-die- gruene-festung, 10810590,10631220.html, last accessed 04.09.2014

Senate Department for Urban Development and the Environment (ed.):
Kreuzberg/Oranienstarße Centre,
http://www.stadtentwicklung.berlin.de/wohnen/quartiersmanagem

ent/en/nkz/index.shtml, last accessed 21.08.2014

Senate Department for Urban Development and the Environment (ed.): 23rd Report on Urban Renewal, 01.01.2000 - 31.12.2001, http://www.stadtentwicklung.berlin.de/staedtebau/foerderprogramme/stadterneuerung/de/download/23StB-00.pdf, last accessed 23.08.2014

Senate Department for Urban Development and the Environment (ed.):
Geodata catalogue / Soldner Berlin Y: 18586.11 X: 27215.53, Foreigners 2008 (LOR) - Turkish citizens, http://fbinter.stadt-berlin.de/fb/index.jsp, last accessed 27.08.2014

Senate Department for Urban Development and the Environment (ed.):
8th Kreuzberg centre / Wassertorplatz, http://www.stadtentwicklung.berlin.de/wohnen/quartiersmanagement/en/evaluation/download/1044_download.pdf, last accessed 27/08/2014

Senate Department for Urban Development and the Environment (ed.):
Berlin, Integrated Neighbourhood Development (Socially Integrative City), Integrated Neighbourhood Development Concept for the Kreuzberg - North-East Action Area, http://www.stadtentwicklung.berlin.de/soziale_stadt/aktionsraeume_plus/download/INSEK_Kreuzberg-Nordost_Internetklein.pdf, last

accessed 29/08/2014

Senate Department for Urban Development and the Environment (ed.):
Support programmes and strategies for vibrant neighbourhoods, http://www.stadtentwicklung.berlin.de/wohnen/quartiersmanagem ent/, last accessed 29.08.2014

Senate Department for Urban Development and the Environment (ed.):
Kreuzberg/Oranienstraße centre, http://www.stadtentwicklung.berlin.de/wohnen/quartiersmanagem ent/en/nkz/index.shtml, last accessed 29/08/2014

Senate Department for Urban Development (ed.): Integriertes Statdteilentwicklungskonzept, Aktionsraum Kreuzberg-Nordost, May 2012, http://www.stadtentwicklung.berlin.de/soziale_stadt/aktionsraeum e_plus/download/INSEK_Aplus_Kreuzberg_2012.pdf, last accessed 10/09/2014

Senate Department for Urban Development and the Environment (ed.):
Handbook on participation - Participation offers opportunities, http://www.stadtentwicklung.berlin.de/soziale_stadt/partizipation/d ownload/Handbuch_Partizipation.pdf, last accessed 16/09/2014

Yücel, Deniz: For the love of Allah,

http://www.trend.infopartisan.net/trd0105/t360105.html, last accessed 04/09/2014

Archive directory

Friedrichshain-Kreuzberg District Office of Berlin, Urban Development Office, Building Inspectorate; building file archive for Kottbusser Straße 8

Illustrations

Cover picture: Türk Discount - Schultheiss Stübchen, own illustration, 12.09.2012

Fig. 1: Turkish "guest workers" arriving at Munich Central Station, early 1960, period Bavaria, http://archiv.deutsch-tuerkische-nachrichten.de/wp- content/uploads/2011/01/foto-Karakoyun-DTN3-600x437.jpg, last accessed 11/09/2014

Fig. 2: Ideal type of oriental city, Spektrum der Wissenschaft, http://www.spektrum.de/lexika/images/geogr/isla_or1_w.jpg, last accessed 13 August 2013

Fig. 3: Kottbusser Tor redevelopment area, http://www.sanierung-berlin.de/sanberlin/hauptteil_sanierung_in_den_bez/sg_san_kottb _tor.jpg, last accessed 11/09/2014

Fig. 4: Overview map of Kreuzberg centre / Oranienstraße, http://www.quartiersmanagement-berlin.de/typo3temp/pics/883ef19336.jpg, last accessed 11.09.2014

Fig. 5: Mission statement "Building bridges - opening gates - strengthening networks", http://www.stadtentwicklung.berlin.de/soziale_stadt/aktionsraeume_plus/download/INSEK_Aplus_Kreuzberg_2012.pdf, last accessed 11/09/2014

Fig. 6: Celalettin Kesim memorial, own illustration, 12/09/2014

Fig. 7: Kottbusser Straße 8, own illustration, 12/09/2014

Fig. 8: Sevim Celebi-Gottschlich, http://www.bpb.de/cache/images/0/43200-3x2- article620.jpg?B1025, last accessed 11/09/2014

Fig. 9: Garden of the Kreuzberg Museum, own illustration, 12/09/2014
Fig. 10: Memorial plaque in the front garden of the Aziz Nesin primary school, own illustration, 12/09/2014

FSC
www.fsc.org
MIX
Papier | Fördert
gute Waldnutzung
FSC® C083411

Kolibri 360 GmbH
Ferdinand-Jühlke-Straße 7
99095 Erfurt, Deutschland
produktsicherheit@kolibri360.de

Druck:
CPI Druckdienstleistungen GmbH
im Auftrag der
Kolibri 360 GmbH
Ferdinand-Jühlke-Str. 7
99095 Erfurt